THE INFOMERCIAL
&
DRTV HANDBOOK

Written By
DREW CUMMINGS

WHAT IS AN "IN-FO-MER-CIAL"?

The definition of an infomercial has changed over the years, and the change now includes both long form (30 minutes) and short form (30 to 120 seconds) television offer for a product or service by consumer marketers to create product awareness, drive retail sales, generate leads, or sell a product or service to the consumer directly utilizing television as the medium and telephone operators as the order takers. The television viewer may order the product or receive additional information by phoning a toll free number which is manned by an inbound telemarketing service. The format of the Infomercial can be right to the point where there is no mistaking that the viewer is being offered the product by the program, or an infomercial can be cleverly produced to resemble an entertainment television format, talk show or reality program. The program gives in depth information about the product or service, and is a long form paid commercial advertisement.

Infomercials proliferated in the United States after 1984 when the Federal Communications Commission (FCC) eliminated regulations that were established in the 1950s and 1960s to govern the commercial content of television. Infomercials particularly exploded in the mid-1990s with motivational products, personal development products, and infamous "get-rich-quick schemes based on the premise that one could quickly become wealthy by either selling anything through classified ads or through real estate flipping. These were hawked by personalities such as Don Lapre and Carleton H. Sheets, among others.

DEDICATED TO

MY WIFE TAMMY, AND MY SONS ANDREW AND BRANDON. THANK YOU FOR YOUR LOVE AND SUPPORT

Chapter 1
OVERVIEW

As a writer, producer, and director in the television industry over the last 30 years, I have observed many changes in broadcast television and the proliferation of cable networks and the internet. But no form of direct marketing using television has grown so quickly, or is so alluring than that of direct response television (DRTV). Direct Response Television is a generic term which encompasses Infomercials, home shopping television (electronic retailing), and direct response spots (short form). Although this handbook will be geared mainly to the subject of Infomercials, the principles and philosophy remain constant for all direct response television. It doesn't matter if the direct response television offer is two minutes or thirty minutes, the ultimate goal of the consumer marketer is to get the consumer to pick up their phones to either learn more about the product or service being offered, drive retail sales, or order the product directly from the television offer. In 1995 when this book was first published. Industry estimates of direct response television sales were expected to exceed $2 billion dollars. In 2009, infomercial sales accounted for revenue in excess of $150 billion dollars.

As a young producer and director in New York City during the early seventies, I produced and directed many DRTV commercials.   These spots were two to three minutes in length, and were always tied into a theme (i.e.: The Greatest Songs of the 50's).   The budgets were extremely low ($1,500 to $3,000), and in some cases the recordings being offered were performed by "sound alike" artists, not even the original groups or singers. Nonetheless, these low budget DRTV spots were responsible for thousands of orders and hundreds of thousands of dollars in sales through short form direct response television offers during the early 70's.

I continued producing these types of direct response commercials during the early 70's on a variety of products.   These products ranged from Kenny Rogers "Quick Pickin Fun Strummin Guitar Course" to the "ExecCycle" and numerous recording artist collections

These short form direct response marketing commercials were the beginnings of a multi-billion dollar industry and spawn a longer form which would become known today as Infomercials.

Twenty years later, after writing, producing and directing traditional entertainment, comedy, documentary, and reality shows for network, first run syndication, home video, cable, and winning numerous awards over the years, little did I know I would come full circle and be drawn back to adding direct response television to the types of programming I would come involved with during the late 80's and early 90's. There is something about writing, producing, directing, and planning a direct response campaign that is even more rewarding in many respects than having a show you have produced being aired on National television. There is instant feedback and gratification when the phones start ringing and consumers purchase the product or service being offered. You know almost immediately if your direct response campaign is working. Either the phones ring or they don't. It's as simple as that!

I first became involved with long form direct response television (electronic retailing) in 1986, as the creator, producer, and packager of the first upscale home shopping show broadcast television. Our partners were Dayton-Hudson Department Stores which are headquartered in Minneapolis. At the time, Dayton-Hudson was the second largest retail chain in America, owning not only Dayton-Hudson stores throughout the Midwest, but also the Target retail store chain.

The show was called "DH-DIRECT". It was an hour long scripted show which aired five days per week. The show was cleared in the top twenty markets in the U.S. as a 13 week test. From a creative standpoint, the show was first class. Elaborate sets were designed to create the look and feel of a shopping mall, with different departments for jewelry, apparel, electronics, house wares, etc. In addition to four regular hosts, fourteen of the top models from Elite Agency were hired to help demonstrate the products on stage and on location. Two location segments were also produced for each one hour show. **A total of 65 one hour shows and 130 location segments were produced.**

Celebrity guests, manufacturer's representatives and famous designers were also booked on the show. To add additional credibility to the show, a team of on-air expects in the areas of electronics, cosmetics, fashion, jewelry, and cooking were also featured on each show. In retrospect, the production and creation of this series placed us as one of the pioneers of "home shopping & electronic retailing", a fact of which I am very proud.

DH-DIRECT emulated the atmosphere of being in a Dayton-Hudson store, known for its impeccable merchandising of upscale product, great "name brand" selections, and high profile atmosphere. We successfully managed to capture the look of Dayton-Hudson in the production of the series, and the executives at Dayton-Hudson as well as the manufactures were very pleased with the look and presentation of their merchandise.

From that point on was where the problems began. I told the executives at Dayton-Hudson that although the show looked great, and Dayton-Hudson was a credible name in retailing, that price points, perceived value, and product quality would be the key to success. "There also needed to be an urgency factor or greed factor built into the show". I went on to say that "You can't sell upscale or designer label products for full retail price points without adding a "gift with purchase" or bundling the merchandise with another product, and those were the elements that would get the consumer to pick up their phones to order.

Dayton's position was they were trying to emulate their department stores, and that they did not want to take the chance of compromising their image as a quality store by offering merchandise at discount prices or adding any of the necessary bundling or gift-with-purchase elements that we felt were necessary for a successful direct response series.

I understood their concerns, however, that should have been discussed before they committed millions of dollars to the project. With those restrictions, I knew that a multi-million dollar flop was forthcoming. The show went on the air for thirteen weeks (65 hours), and at the end, Dayton-Hudson had a lot of red ink on the books, and other networks like QVC and HSN that were about to launch learned some very valuable lessons studying our series.

Since that time, however, home shopping networks and electronic retailing have proven to be very successful, and consumer has become more comfortable ordering products from television. Networks such as QVC and HSN will generate hundreds of millions of dollars in sales by 1994, and in the 21st century, sales are in the billions.

Both of these electronic retail networks have been responsible for creating that consumer comfort in ordering merchandise from television over the last seven years. It's only a matter of time before retailers such as Macy's, Bullock's, Neiman Marcus, and the major catalogue retailers will successfully develop and launch a new generation of upscale electronic retailing networks and programming that will incorporate many of the high quality production values, entertainment values, and information incorporated into DH-DIRECT.

I mention the Dayton- Hudson project, because home shopping shows, two minutes direct response commercials, and what has become known today as Infomercials, can all be categorized as direct response television (DRTV).

Infomercials are not a new form of direct response marketing for television. They have been around for twenty years in one form or another, such as the two and three minute commercials selling everything from sun glasses to record album offers. Even religious programming is defined as direct response, where the minister or preacher asks for donations from the television viewer.

One of the most successful two minute Infomercials over the years was the "Garden Weasel", a mechanical garden divide that extracts weeds by the roots with minimal amount of effort. With reported sales of over one million units, this is an early excellent example of a simple product that met all the criteria for a successful direct response campaign. The "Garden Weasel" subsequently found its way to retail stores throughout the world, and is a further example of how a direct response television campaign can introduce, advertise, and ultimately drive a product to retail.

If you were to speak with industry experts, producers, T.V. station program directors, or advertising agencies about Infomercials or direct response television in the early 90's, they would all have told you that Infomercials were for the most part, a low budget, classless form of marketing and advertising fraught with "snake oil" salesmen trying to bilk the American public from their hard earned dollars on worthless products.

In many instances they were correct. However, as with all growing industries, those who chose to get rich quick with untrue claims and inferior products are either in jail, have straightened up their acts, or are out of the business. Taking their place are entrepreneurs who are talented marketing geniuses and creative producers who can identify a good product and market it in the form of an Infomercial.

Many of the "new breed" of Infomercial writers and producers come from the print side of "direct response", where they honed their skills selling products via newspaper and major publications.

Hollywood has also jumped on the bandwagon, with many top producers of traditional entertainment programming vying for Infomercial production dollars from consumer marketers and manufacturers with their advertising agencies handling the Infomercial campaigns.

Today, Infomercials have proven to be a new and effective means of marketing, selling, and driving retail sales of a product or service. Respected celebrities and production companies are now endorsing and producing Infomercials for a wide variety of products and services.

When this book was originally published in 1995, artists such as John Ritter, Lisa Hartman, Ali McGraw, John Davidson, Fran Tarkenton, Cher, Kathy Smith, Victoria Principal, Joe Namath, Dionne Warwick, and Meredith Baxter-Birney are only a few of the many Hollywood celebrities endorsing and hosting Infomercials. Today there is no difference, with top Hollywood and sports celebrities pitching products ranging from self improvement tapes and "Get Rich" real estate systems to cosmetics and medical breakthrough treatments. The production values of Infomercials are also rising with many budgets approaching $350,000.00.

From a broadcaster's standpoint, Infomercials in today's economy are widely accepted by cable networks and commercials broadcasters alike. Why? Because they serve two purposes in a fragmented broadcast economy where ratings and viewers are dropping and cable networks are adding niche channels that fragment the viewership even more. Infomercials provide free programming to stations in addition to generating instant cash flow without the costs of sales. Stations and program directors, who two years ago vowed never air Infomercials, are now first in line to take Infomercial dollars.

From a manufacturer's or consumer marketer's viewpoint, Infomercials are a cost effective means of introducing, advertising and selling products or services, which cannot be easily demonstrated in a one or two minute commercial. It is also a means of creating consumer awareness on a new product. Infomercials are also very effective at generating qualified leads and driving retail sales.

With Infomercials, the product is presented in a format that gives it more exposure at a lower cost than is possible with more traditional forms of media. The Infomercial also allows the product to be demonstrated in detail over length of the program. It is "The Perfect Salesman"

Chapter 2

THE SUCCESS OF INFOMERCIALS

The successful Infomercial is one that offers a great product at a great price and serves the needs of the masses. There are new products that are introduced all the time, but a few past examples of successful Infomercials are "Victoria Jackson Cosmetics", (American Telecast), "Flying Lure" (Quantum Marketing), Tony Robbins "Personal Power" (Guthy-Renker Productions), Susan Powter's "Stop The Insanity" (USA Direct), and Jay Kordich's "Juiceman" (Trillium Health Products) just to name a few. Victoria Jackson Cosmetics has reported gross sales of approximately $72,000,000.00 per year. John Ritter's "Where There's A Will, There's An "A" (American Telecast) is reported to have grossed well over $100, 000,000, and is still going strong. In fact, all of the above mentioned products have grossed in the hundreds of millions of dollars.

Equally as important is when executed properly, Infomercials serve as a launching pad for product at retail, as well as catalogue or the ever-growing internet direct response market.

There are literally hundreds of products that were initially introduced to the American public through an Infomercial that are now available at retail stores throughout the world. Because of the successful sales and television exposure, the manufacturers were able to bring the product to retail very successfully in National retail chains across America.

The power of Infomercials in politics was introduced in the fall of 1992 with Presidential candidate Ross Perot. It is estimated that over 20 million viewers watched Mr. Perot's 30 minute Infomercial. This effective use of Infomercials demonstrated to the consumer marketer that long form advertising could communicate a message in a way no other form of marketing or promotion could attain.

The success of Ross Perot's campaign using Infomercials has prompted many Fortune 500 marketers to produce Infomercials. Just a few of these companies preparing or utilizing long form direct response are

Kodak, Avon, Philips, Electronics, Volkswagen, Redken Laboratories, BMW, Range Rover, Braun, Black & Decker, GTE, Club Med, Bissell, and Corning Vitro just to name a few.

Chapter 3

CHOICE OF PRODUCT

Virtually any product that appeals to the mass public, can be demonstrated easily and serve a functional purpose in the consumers mind can be incorporated into an Infomercial. This doesn't mean just because someone came up with a new "mouse trap", they should go out and produce an Infomercial. Pick a product with wide general public appeal, not a specialty product attracting only a small target audience. When choosing a product, also make sure that you can get quick delivery and ample supply from the manufacturer.

An early pioneer in the infomercial business, Quantum Marketing, fell into that dilemma when they were producing and running their "Vertical Roaster" Infomercial. The product was selling like "hotcakes"; however, the demand was greater than the supply. They just couldn't get enough product from the Orient to fulfill the orders, thus, their media costs and cost per order ran extremely high. The point here is to make sure that you have access to an abundant supply of product.

Not that you have to warehouse a large inventory, just know how soon it takes for the manufacturing and delivery process. As more and more consumer marketers utilize the Infomercial to drive retail sales or introduce a product or service to retail, the same holds true. Make sure the product is in the stores.

Why do people buy from Infomercials? Research has proven successful Infomercials fall into three categories. From a psychological standpoint the consumer will pick up the phone and order from an Infomercial for the following reasons:

- **GREED:** People want to get something for nothing. It's a proven fact. Show someone a product that fulfills a need in their life, or creates a perceived need in their mind, and offer it at a price point they perceive to be a great value, and they will pick up their phones to order every time. Create the illusion they are receiving something for nothing, as in a free gift, or additional product for "free", and you add even more odds the public will pick up their phones or visit their local retailer. No one did this better than Mike Levy in the late 90's on his Infomercial series "Amazing Discoveries". Like it or not, "Amazing Discoveries" and Levy are responsible for probably more product sales during that time using Infomercials than any other company in the world.

♦ It is estimated that close to $500,000,000.00 have been spent by consumers over the years, buying product from Levy's over enthusiastic pitch. Why are they so successful? Because they found products that fit into the three basic criteria for a successful Infomercial. And most of the time the product or service has more than one of the three basic success elements incorporated. And in the future look for more manufacturers and marketers to use Infomercials to drive retail sales on television as well as the internet.

♦ **VANITY:** Create the illusion that a product will make someone look year's younger, change their physical appearance for the better, or just makes them feel good about them and improve their quality of life, and you have a product with a good chance of succeeding. Let's take Victoria Jackson and her highly successful cosmetic line and Infomercial of the 90's. With reported sales exceeding 10,000 orders per week at an estimated $150.00 per order you can see how the vanity factor can really pay off.

Similar Infomercials that have been extremely successful is Cher's hair product Infomercial, Victoria Principal cosmetics, Nordic Track system, and the many hair products used to cover baldness or whiten teeth.

♦ Yes, the cosmetic line is a good line, but there are other cosmetic lines equally as good that are available at retail stores throughout the world at competitive pricing. Why then are the women of America picking up their phones at the rate of 10,000 per week?

- It's simple! Vanity and credibility. Victoria Jackson is a make-up artist and cosmetologist to the stars, and endorsements and testimonials from major celebrities who are perceived by the public to be some of the most beautiful women in Hollywood are pitching her line of cosmetics. The subliminal message to the public on Victoria Jackson's Infomercial or any other beauty or health product offered by Infomercial's is simple and basic. "Buy my product and look like a Hollywood movie star". The "before and after" demonstrations or make-overs using everyday women are also an extremely effective visual image and sales tool.

- **URGENCY:** All types of methods are incorporated into successful Infomercials to accomplish this. One of the most successful means of creating urgency is to tell consumer "This is not available in stores and is available only through this television offer for a limited time, or while supplies last". This is the oldest and probably the most effective means of getting people to pick up their phone. It is an element that should be incorporated into every Infomercial if possible.

♦ Remember, instant gratification and impulse buying also fall under this category. So try to create an urgency for people to pick up their phones. In conclusion, you can have the greatest product since the beginning of time, however, if you do not incorporate at least two or more of the three main elements into your Infomercial, it will never be a success. What works? Well, I have mentioned a variety of products that have successfully worked on Infomercials and home shopping. Usually, an Infomercial deals with one product only. However, there is a hybrid Infomercial form cropping up across America, and will be utilized in the coming years as media prices escalate. It is a cross between a traditional Infomercial and home shopping show which offers multiple related products on one show.

I recently produced such an Infomercial for a manufacturer of children's products which ran during the fourth quarter of 1991. There were five products in the Infomercial, all of which could be used in a child's bathroom. The objective of the Infomercial was to drive retail stores, as the products were already in stores such as Toys 'R' Us. The product was offered over the air via an 800 number, however, the prices offered on television were higher than retail.

The theory was that if the consumer wanted to order from television, great, but the prime objective was to drive them to retail. The campaign was very successful, with over 100, 000 units sold at retail across America in the fourth quarter of 1991, with an average price point of $39.95.

The manufacturer was happy because they made some television sales as well as supported retail sales. The retailer was happy because the product was advertised, and their retail price points were protected.

As I mention throughout this book, you will see more and more Infomercials being produced in the months and years to come, which will be designed to drive retail sales or introduce a new product to retail, with less emphasis on television sales.

Chapter 4

THE POWER OF TESTIMONIALS

Show me someone who looks like my next door neighbor giving praise to a product, or telling me how it changed his life, made her look better or feel better, helped him earn more money to buy the house, boat, or yacht, and you'll have pull me away from the phone. Outside of the three emotional elements that go into successful Infomercial formula, there is one more key element that is mandatory…. The Testimonial. Even more potent than a Hollywood celebrity endorsement who has used the product and met with unbelievable succes

Of course, if you're planning to produce an Infomercial next week, you had better have at least six to ten average people who have tried the product over an extended period of time, and are willing to be taped or filmed giving their praise. For these are the people that will bridge the

gap between the host, who is recognized as the pitch person selling the product, and the general public.

Nothing is more believable than to hear it from people like you and me who have used and tested the product, and give testimonials as to how great it is, or how it has changed their life. But whatever you do, make sure that you do not pay these people a fee for their endorsements, or script what they will be saying. This is a "no, no", and the FCC, FTC, and or Attorney General are there to watch for such deceitful tactics. Make no mistake, they will pull you off the air and prosecute you to the highest extent of the Law. If you find that you are in a situation where you need to pay for testimonials, make sure you place a visual disclaimer on the Infomercial notifying the consumer that people endorsing the product have been compensated. Remember, with pre-planning, and "real" people praising the product, there is nothing more essential than testimonials as the key selling element of any successful Infomercial. It is also very important when taping your "man on the street" to use people that look like everyday people. Find a good mix of ethnic backgrounds, different sizes, ages, and shapes.

If you have the budget, additional endorsements from "expects" or celebrities who have used the product or service and have a high recognizable factor that people feel they can trust, is a wise investment. John Ritter and the late Michael Landon are only two such celebrities that have hosted and endorsed very successful Infomercials.

Chapter 5

THE PITCH MAN OR WOMAN

A believable pitch person, or host of the Infomercial is the next element that must carefully be considered. The pitch person and the host can be two different people, as in "Mike Levy's, Amazing Discoveries Infomercial series. Levy acts as the Host, and the person who either represents the manufacturer or who invented the product or service is the pitch person. Levy poses as the disbeliever who needs more proof on a product's claims, and is also the person who supposedly acts on behalf of the consumer when the pricing of the product is mentioned.

There are also very successful Infomercials where the pitch person and the host are one in the same. The first decade of the 21st century saw probably the most successful pitch person in the history of direct response television. Billy May, who honed his pitch talent on the boardwalk of Atlantic City, was probably responsible for more product sales during his short career than any pitchman or infomercial host in the history of direct response television. His untimely death in 2009 left an impressive legacy that future hosts and pitchmen and women will aspire to emulate for years to come.

Over the years, products such as the sandwich maker, the vertical roaster, the wok, and the food saver have all made millions of dollars in direct response sales, and all without the help of a notable celebrity. Why you ask? It is simple. All of the above mentioned products serve a very functional use in the home, are available at great pricing, and are pitched by people who have been doing it for years.

These hosts/pitch persons, without exception, have honed their pitches at swap meets, county fairs, and seminars across the nation for years. The sandwich maker is not a new product that just happened to take America by storm.

That product has been available from county fair salesmen for fifteen or twenty years. It was, however, the Infomercial that made the sandwich makers and some of the other products mentioned the hottest selling house ware products in America. Products pitched by these people who have fine tuned their sales pitch way before the word Infomercial was invented. All they had to do was translate their pitch to television.

It is these hosts/pitch men and women who know how to hype the product, answer questions and solve problems the consumer may have without being asked, create a perceived urgency for the consumer to pick up the phone, demonstrate every conceivable use and application of the product, and then "ask for the order", or in Infomercial terms "Call To Action".

Every successful Infomercial MUST have either the Host or Pitch Person "ask" for the order, or the "Call To Action" as part of the show format if the Infomercial is designed to sell direct from television. DO NOT leave it up to the voice-over announcer who gives the "here's how to order" pitch to be your final salesman, it usually doesn't work.

Then there are the types of products that promise wealth from following a self improvement course as, a Real Estate course, or the highly successful personal motivational courses such as Anthony Robbins' "Personal Power" course produced by Guthy-Renker and hosted by Fran Tarkenton. in the 90's Anthony Robbins reportedly sold 250 million dollars worth of his motivational courses via the Infomercial. Robbins has fine tuned his course over many years at private seminars across the country. If there is one example of a perfect Infomercial that sells a product or service that encompasses all of the three key elements, as well as high level testimonials from celebrities, judges, sports stars, as well as the guy next door, this Infomercial should be used as the template.

In addition, Fran Tarkenton, one of sports most respected athletes of the century, is hosting the show, in addition to endorsing the product and serving along with Robbins as the pitch person who gets the consumer to pick up the phone.

Let's take another example of the perfect pitch person. USA Direct reportedly spent approximately $30,000 on the production of Susan Powter's "Stop The Insanity" in the 90's, one of the most successful Infomercials in history, selling approximately 15,000 units per week.

How were they able to produce this Infomercial so inexpensively? Without taking credit away from USA Direct, who decided to do the project after it had been unsuccessfully pitched to a number of other Infomercial production companies, they produced and packaged a program which looked much more expensive than its budget. Keep in mind all they did was merely capture what Susan Powter does for a living.... sell her diet program and video tapes.

Susan Powter has been fine tuning her pitch for years in smaller venues than depicted in the Infomercial. She has covered the county giving seminars on health, weight, and exercise for almost ten years. All the producers of her Infomercial did was to capture her pitch and performance on video tape in a larger arena and with a larger audience than she normally works.

The star of the Infomercial is Susan Powter, the product "is" Susan Powter, and the pitch could only be given by Susan Powter.

And what can you expect to pay for such a pitch person or celebrity host? If this were in the early years of Infomercials, and if you could find a celebrity who would take the chance of jeopardizing their credibility to host an Infomercial, you would probably have paid a flat fee of anywhere from $25,000.00 to $35,000.00.

As little as two years ago, no respectful talent agency in Hollywood would even think of sending a celebrity on a casting call for an Infomercial, unless the celebrity's career was in a slump, or the celebrity just needed the work to make ends meet.

Today, all that has changed. With reported royalty fees for celebrities approaching seven figures annually for their hosting chores, all of the major talent agencies have formed Infomercial divisions to handle the demand for top names in the entertainment industry to host a variety of Infomercials.

It was the early successes that pushed talent fees through the ceiling, not to mention profit participation. If you were to hire a celebrity of any measurable stature as a spokesperson to host an Infomercial in today's market, you can expect to pay in the range of $100,000.00 to $250,000.00, plus an average royalty of between 1% to 3% of the gross television sales, and a small percentage (.0025%) of the retail sales.

The choice of whether or not to use a celebrity spokesperson lies in the nature of the product. If the product is able to be easily demonstrated as in kitchen products, it may not be necessary to hire a celebrity spokesperson.

However, if the product being offered is an intangible, such as a weight loss program, beauty and cosmetics, or self improvement course, it is advisable to have a credible third party endorsement from a top celebrity represent the product.

If you are planning to produce an Infomercial utilizing the talents of a celebrity, it is advisable to hire a production company or consultant that deals with celebrities on a daily basis. These production companies have relationships with the top agencies, managers, or celebrities, and know how to negotiate the best deals. More importantly, they know how to work with celebrity talent.

Also, in most cases, don't expect to finalize negotiations on a celebrity spokesperson in a few hours or days.

Negotiating for talent takes time and patience. Back and forth offers with agents and managers, time conflicts, and turndowns can be a time consuming and lengthy process. Another thing to remember in dealing with celebrity talent to represent your product on an Infomercial is that you need time in which the celebrity gets to know the product "inside and out".

Don't assume that just because you have hired a professional spokesperson they will come on the set or location on filming day and act their way through the production. Rehearsal time is essential.

Work with the talent for weeks in advance of taping the Infomercial. Let them become comfortable with the product. They should know every aspect of the product, have handled it and worked with the product, and be able to look and feel comfortable with the pitch and message you want to get across. Remember, in most cases, celebrities are not familiar with the Infomercial format.

Their whole careers as actors on television, film, and commercials have been carefully scripted. Many of these celebrities may not be comfortable conducting interviews, or speaking directly to the camera. There is nothing more devastating to an Infomercial than to have a celebrity spokesperson look uninformed or uncomfortable.

In choosing a celebrity, try to find one that has had a personal experience that can relate to your product and who believes in the pitch. It may be a waste of money to hire a celebrity whose only function is to hype the product, but has no personal experience to add to the pitch.

Chapter 6

COST OF PRODUCT

As a rule of thumb, the cost of manufacturing the product being offered on Infomercials should be no less than five times the selling price. There are also infomercials where the ratio is as high as ten to fifteen times the selling price. The pricing of your product on an Infomercial is a major factor. Research shows that $14.95, $19.95, and $29.95 are magic numbers. For whatever reason, $24.95, and $39.95 seem to be price points that tend to be unsuccessful. Pricing also becomes a key issue when a product is being sold for over $49.95.

A multiple payment plan should be offered to the consumer when pricing is $59.95 and above.

Just keep in mind that by Law with multiple payment plans, the product must be shipped to the consumer upon credit card approval or receipt of the first payment. Subsequent payments for the product by credit card after the consumer has the product must be separately cleared on a monthly basis until the full selling price is received. Therefore, it is imperative the first payment received for the product cover the cost of manufacturing plus shipping and handling. Remember, the consumer already has the product, and there are no guarantees subsequent payments will be approved on the credit card. In fact, you should plan on not receiving complete payment on a small percentage of orders.

There are exceptions to every rule, however, when it comes to pricing a product for an Infomercial, and the costs of that product. There is a new form of Infomercial emerging slowly through the glut of self improvement products, get rich quick courses and products promising to change the quality of life. The goal of these Infomercials is cloaked within a "hidden agenda".... RETAIL. Yes, these new Infomercials will look and feel like the traditional fare, but the "hidden agenda" is to eventually move the product into main stream retail stores or enhance sales for an existing product already at retail. Where else, but in an Infomercial, can you introduce a product for half hour and make a possible sale on the air via direct response.

Furthermore, a manufacturer or distributor can also promise advertising support to potential retailers, that if they order the product they will be supported by additional advertising in the form of the Infomercial.

With this type of long term "hidden agenda", the cost of product does not have to be the three to five times the selling price. The profit margin derived strictly from the Infomercial sales becomes less important if the product is headed for retail. In some cases, there are never sales directly from television, only leads are generated.

In this scenario, the Infomercial serves as a launching pad to support retail sales. In the months to come, you will also see Fortune 500 companies begin to produce and air another hybrid form of an Infomercial, pitching everything from cars and health care plans, to high tech electronics, diet plans, and vacation getaways.

These Infomercials, however, most likely will not offer the product over the air. They will be designed strictly to create consumer awareness and generate "leads".

Research by many household appliance manufacturers suggest by the year 2010, 90% of all Infomercials will be used to generate sales leads, drive retail sales, or create public awareness.

I produced such an Infomercial which was designed to introduce a new product to the American public. The product was called "Aqua Cleanse". It was developed in Japan, and is a total skin care cleansing system that works off of water pressure from the shower or sink. When the North American distributor came to me to produce the Infomercial, the product had never been seen in the U.S. The distributor wanted an Infomercial that could serve both as an advertising vehicle and demonstration to introduce the product to both the retailers and the general public.

The product was offered to the public via a toll free number at $149.95, and although the sales from the Infomercial itself were less than impressive, the distributor was able to secure sales in major department stores throughout the county where they retailed the product at $119.95.

The Infomercial by its design was successful in launching and supporting the product into retail. The ultimate icing on the cake would be with an Infomercial offering a product both on television and at retail, and the Infomercial would generate enough telephone orders at the higher price point to pay for media costs.

The problem in introducing Infomercial products to retail in the past is that many of the products sold on Infomercials have not been accepted by main stream retailers such as Walmart and Sears, in terms of image or quality. Many of these large retail chains have been burned in the past with product that has not lived up to the great image portrayed on the Infomercial. They have been plagued with returns and massive customer service complaints. That concern is changing slowly at the retail level now that Infomercials are proving to be an alternative and successful means of launching and advertising a product.

Major manufacturers such as Black & Decker, Singer, Orek, Dyson, and Braun are all considering or have used

Infomercials as launching pads for new products, in addition to products that need more visible demonstration which cannot be accomplished by traditional advertising methods or in-store demonstrations.

Chapter 7

THE SHOW PRODUCTION

As the Infomercial form has grown over recent years, so has the production budgets. The days of $10,000 production costs for a half-hour infomercial do not exist any longer. As Infomercials have evolved, so has the quality of production, and with it, a major increase in production costs. Standards are being set with every new successful Infomercial production. The consumer is also expecting better production values from Infomercials, even though they consciously may not be aware of their changing standards.

If you are a consumer marketer, manufacturer, advertising agency, or entrepreneur looking to produce an Infomercial to sell the product to sell the product or introduce the produce to the marketplace and drive retail, seriously consider "NOT" doing it yourself.

Unless you have experience in producing Infomercials or television shows, leave it to the professionals. That is not to say you should give "carte blanche" to a producer, but you should choose a production company that can take care of the hundreds of elements, both creative and technical, that goes into the production of an Infomercial. The production qualities incorporated into Infomercials have set new standards for Producers.

Great sets, high quality graphics, multiple cameras, creative lighting, original music, professional direction, and creative writing are all key elements incorporated into the two Infomercials mentioned above. The production values that must be incorporated into today's Infomercials are on parallel with network quality productions. Celebrities and agents have also played an integral part in the evolution of production quality.

These celebrities and their agents have not only demanded higher fees, but have also kept a close eye on who is producing the Infomercial, and how much of a budget is contemplated.

The American public, with their exposure to network, cable, and higher priced Infomercial productions, have taken their dollars and spent them on products that are professionally presented and packaged.

They are associating good products with that their eyes have been trained to recognize as high quality production values. They may not realize it, but it is true nonetheless. What should the production costs be on Infomercials? Beware of production companies promising a great production at "bargain basement prices". Choose a production company based on Infomercials they have produced before. Ask to see their previous work. And above all, do not hire a production company based on work they have produced for another form of television such as dramatic, variety, reality, or industrials.

Past credits from another form of television production, no matter how impressive they may be, does not qualify a production company to produce Infomercials or direct response television. The production of Infomercials is a highly sophisticated form of television. It incorporates not only entertainment values, but above all must "SELL" the product.

For this to be accomplished, you must seek production companies that have had past experience producing direct response television Infomercials or DRTV spots.

There are certain "key" elements and "buzz" words that must be incorporated into the script, as well as in the way an Infomercial is photographed and produced. With the morality rate of one success for every eight Infomercials produced, you will need all the experience you can get to produce a successful Infomercial.

Remember, no matter what production company you contact to produce your Infomercial, beware of those who profess to have produced nothing but successful Infomercials. I guarantee you that they have produced more failures than successes.

Ultimately, you must trust your eyes, your ears, references, and industry reports and surveys. One of the most respected industry reports is the Jordan Whitney Report, which analyzes over 200 Infomercials on a weekly basis. To learn more about this report contact the Jordan Whitney Report, Inc. at (714) 832-3353, or visit their website at:

http://www.jwgreensheet.com/consulting.asp.

If you see an Infomercial on the air that looks great, expect to pay a royalty to the producer and writer or production company of an Infomercial should you choose to hire a production company. An average combined aggregate royalty should be approximately 3% to 5% of the gross television sales.

Your production costs will vary with your choice of hosts, number of location days require (if any), number of cameras, costs of set construction, etc. There is no such thing as a typical production. Each one has their differences.

To try to arrive at a mid-range production cost, figure on spending a minimum of $50,000.00 for an acceptable quality production with very few frills. The intangibles that will raise the production costs will be set costs, producer, writer, director, talent, and how elaborate post production editing will be.

If you want all the "bells and whistles", with state of the art special visual effects, then you can expect to pay upwards of $350,000.00 for a half hour Infomercial. Today, most infomercials (DRTV) are produced in either SD (standard Definition) or HD (High Definition), depending on where they show will be broadcast.

Chapter 8

THE SHOW FORMAT

The format of the Infomercial is broken down into three segments, with a variety of elements within each segment. I will try to take each element of the show and break it down into simple terms.

THE SCRIPT: Do not think of the script for an Infomercial in the same terms as a script for a movie or dramatic series for television. It is more like a talk show such as Oprah. Talk show scripts, as with Infomercial scripts, are much more impromptu. When Oprah does a show, her staff has done some basic research on the guests, has a background on the subject matter of the day, and knows the direction which she wants to take the show. Every word that Oprah says is not on a cue card or teleprompter. However, she does have the "bullet" points and what is called "cheat" notes available to her, either written down paper in her hand, or on a teleprompter. These notes are information and key elements that must be discussed during her show. This is similar to the type of "script" used in Infomercials.

Not every word is written down for the host to memorize, but key points concerning the product that must be discussed during the course of the show. There are exceptions to this rule, and some Infomercials have used entirely scripted material.

Therefore, it is essential that the Host or Pitch person you choose to host the Infomercial feels comfortable with this type of format. He or she should have the background, if possible, at being able to conduct an interview show, or have credits showing they are able to improvise in this type of setting. I have found that the perfect pitch person is one who either invented the product or has sold the product at county fairs or conducted seminars for many years. The host should also have on-camera experience and above all, a complete and thorough familiarity with the product or service that he or she is selling.

The script then, is merely a series of "bullet" points that give an outline or establish certain key points that must be discussed or pitched during the course of the show. The format should be entertaining, as well as informative.

There is one very important factor that goes into the script and format of an Infomercial. Research has shown we are in a generation of television viewers that have "remote-control-itis", so most television viewers watching television are forever changing their dials throughout the course of their television viewing. This is why it is critical to constantly have the host or hostess repeat the same key selling points of the product pitch throughout the show.

Areas that should always be scripted to the "word" are:

♦ The opening of the Infomercial
♦ The technical and descriptive data
♦ The "commercial" portion of the Infomercial
♦ The "call to action" which is when the host "tells" the consumer to pick up to phone to order.
♦ The information "plate" which gives telephone number, internet website, and ordering information.

Choosing a writer with the knowledge and experience at writing Infomercials or direct response is the key to a successful production. The mistake most novice producers or consumer marketers make is to hire a writer from another medium such as newspaper journalists, commercial writers, ad agency copywriters, novelists, T.V. news writers, or manual writers who write the instruction booklets that come along with the product. The common problem with most scripts written by writers who have never written an Infomercial is that they are either too technical, or too entertaining. Many Infomercials fail because the production companies produce very entertaining shows, but forgot to put the "SELL" in the script and format.

Therefore, try to find a writer who has previously written an Infomercial or direct response commercial before. There are only a handful of quality Infomercial writers in the industry, and most of them are working for the major Infomercial production companies. However, many of them are freelance writers who move from assignment to assignment between the handful of Infomercial production companies.

They are hired on a project by project basis, and can be hired by anyone willing to pay the price. If you have trouble finding one of these writers, the best thing to do is to watch the end credits on an Infomercial and note the name of the person who wrote the script. Search their name on the internet using Google, Facebook, or social and business networks. Many writers also have their own websites.

The fee a writer charges can vary from a $5,000.00 flat fee to $10,000.00, depending upon how many successful scripts they have written before, and if those credits put the into a category of a writer who should receive a percentage of the sales. You should not pay any more than $10,000.00 for a half hour script and 3% of gross sales from television if the writer and producer is the same person. Never pay percentages of gross sales to any producer or production company unless you are selling the product or service only through television, or unless the Infomercial is launching a product that has never been offered at retail.

If you feel the producer or production company should be entitled to a percentage of retail in addition to gross TV sales, and acceptable fee would be ¼ to ½ of one percent of wholesale.

THE SETTING: The setting should be an environment on a stage or location that is simple, yet functional. Location shooting is always more expensive than on a stage, due to the logistic and loss of a controlled environment. The setting should allow the product to be easily demonstrated, and the host and guests should be comfortable. It can be as simple as a living room setting with couch, chairs, and product presentation table, or a kitchen setting or the grounds of a luxury hotel or resort

THE "CALL TO ACTION" COMMERCIAL: The word commercial may be misleading when used in context with Infomercials. For the entire Infomercial is one big commercial separated by smaller commercials that segment the show. These commercials are usually one to two minutes in length, and succinctly re-state what viewer has been watching in a more precise, condensed form.

It is here, during the commercial that every claim and pertinent fact the host has presented concerning the product is reinforced and repeated, leading up to **telling** the consumer how to order.

THE PLATE: The plate, also known as the tag, is a visual image showing:

- The name of the product
- The price of the product or payment schedule.
- The 800 number to call.
- The mailing address for viewers to send checks or money orders.
- The credit card emblems accepted.
- The shipping and handling charges.
- The money back or return policy.

In addition to this information being displayed visually, there is an announcer stating the same information. Because most orders from an Infomercial will be through credit card purchases and not checks or money orders, it is suggested that the announcer repeat the Toll Free number numerous times. Another suggestion is that the

Toll Free number be displayed during the body of the show as many times as possible.

Chapter 9
HYBRID INFOMERCIALS

As the direct response industry matures, it is becoming more and more evident that a new breed of Infomercial producer and Infomercial format is also emerging...the **HYBRID INFOMERCIAL**.

These are two types of hybrid formats I would like to discuss in this section. The first type is being produced by a small handful of production companies in the business such as Tyee Productions, Guthy-Renker, Cummings Entertainment, Script To Screen, California Production Group, and American Telecast. By the writing of this book, many more have emerged and a few may not exist, so do your research using the internet.

These hybrid Infomercials are being produced on behalf of Fortune 500 consumer marketers and their advertising agencies. They may be disguised as reality shows, game shows, talk shows, documentaries or special events.

These Infomercials differ in many ways from the traditional Infomercial in that they are designed mainly to create leads, drive retail sales, and introduce or reinforce an existing brand image, while at the same time be entertaining.

They are less "sell" oriented, and they tend to give more information about the company, product, or service bring offered. They also tend to be more entertaining and most important there is not a royalty position for the producer or production company as with traditional Infomercials selling a new product. Trust me, companies such as Volvo, Real Estate developers, or Cruise Lines are not about to pay a royalty for every lead or product sold.

These hybrid Infomercials and their production are being driven by the demand of major consumer manufacturers on their advertising agencies. As of the writing of this revised edition, there is still a major reluctance by the advertising community to venture into the world of Infomercials. However, it is an area that every major as agency is looking into.

It is also an area unfamiliar to most agencies; therefore, they are being very cautious and protective when it comes to their clients utilizing this form as a marketing tool.

Ultimately, it is my opinion the clients will win. The agencies will become more familiar, educated and convinced that Infomercials, as a part of an advertising and marketing strategy can deliver more information, generate more sales and leads, and create more public awareness than traditional advertising at a fraction of the cost.

I also predict that within the next two years, the production and media placement of these hybrid Infomercials will be controlled by the advertising community, much as they control the traditional; spot production and placement today. These agencies will also develop direct response (DRTV) divisions of their companies.

The agency's reluctance for the most part is well founded.

Although production values have increased with the current crop of Infomercials, with the exception of a few they still do not approach the budgets or have the high production values of most national commercials on the air which being produced. Most of today's traditional advertising spots (30 or 60 seconds) and their production budgets can range from $250,000 to $1,000,000.00 for a single 60 second spot.

It is the agencies concern that by utilizing the current Infomercial production standards to augment their clients advertising and marketing dollars, they will compromise their client's image with sub-standard production values.

As more and more Fortune 500 companies and major advertising agencies begin utilizing hybrid Infomercials, it is my belief as well as other industry leaders, that the production community will be dived into two sectors.

One sector will be the production companies only producing high end-high budget Infomercials on a fee-only basis much like commercial spot production.

The other sector will be the Infomercial production and media placement companies serving the entrepreneur or manufacturer launching new or existing product that will sell only through infomercials and DRTV spots and receive a royalty on every product sold.

But where will these hybrid writers, producers and directors come from? Some agencies will gear up Infomercial divisions that will handle creative and production matters in-house and hire the freelance services of knowledge writers, producer, and directors. Other agencies will reach out to the entertainment television producers and spot commercial production companies.

Let's face it, if you are major advertising agency with a multi-million dollar client wishing to utilize Infomercials, who will you look to first to produce and direct your Infomercial? With the exception of a few Infomercial producers, probably either a commercial production company with a hot director who you have worked with in the past on a winning spot campaign, or a production company from the entertainment side of television with credentials in variety, reality, music, or comedy. Like it or not, initially, this is where I see the advertising agency going for their Infomercial production needs in the near future.

I hope I am wrong! Just because a hot commercial director or Emmy award winning entertainment producer has the acceptable image to the agency and client, does not mean he or she can produce a successful direct response television campaign.

In fact, utilizing the services of these talented people alone, without any of them having knowledge or exposure to DRTV will probably be disastrous, as it has been proven in electronic retailing time and time again. Every time an Infomercial is produced by someone that has no experience in DRTV and it fails the industry as a whole take one step backward. Hopefully there will be more and more production companies who have produced these hybrid Infomercials to answer the demand that is forthcoming.

The second form hybrid Infomercial emerging is the marriage of home shopping (electronic retailing) and Infomercials. Let's first analyze the two existing home shopping networks that exist today. HSN and QVC have the home shopping market concerned. Both offer the consumer great products at great prices 24 hours a day. Their format is straight forward and to the point...sell product! If at the same tie the format can be informative and entertaining, great! But the prime mandate is to sell product. From a creative and programming standpoint, the look and format of QVC and HSN is not acceptable to independent or network station affiliates, or cable networks which are not dedicated to a 24 hour home shopping format.

Network television, first run syndication and cable television is where I feel the second hybrid Infomercial format is being developed, and there is a long list of unsuccessful attempts to prove it. Let's name all the failed attempts so far.

DH-Direct (syndicated), NiteCap (ABC), Joan River's "Can We Shop", and first reports of NBC's Mall Of America is less than promising. In the batters box is Macy TV, Fingerhutt, and a host of other potential

electronic retailers about to get into direct response television and electronic retailing.

If we have learned anything from past failed attempts at bringing long form direct response to television, which will offer multiple products to the consumer, it is that:

- Too much sell and the show looks like QVC or HSN.
- Too much entertainment and there is no product sold.
- Wrong choice of product offered at price points that are too high equals a failure.

Another obstacle to contend with is that in some cases, these failed attempts were also expected to generate ratings for advertisers during commercial breaks.

I feel that in the very near future, a hybrid format that is entertainment, informative, offering great product at great prices will be developed for broadcast television. In my opinion, however, for this hybrid to work it will have to be developed and executes by both the retail and entertainment communities collectively.

Chapter 10

MEDIA & IN-BOUND TELEMARKETING

These two elements (media and in-bound telemarketing) are the reason for the success or failure of an Infomercial. You can have the best product along with the greatest host, and the most professionally produced Infomercial, but if the media or in-bound telemarketing service is not chosen carefully, you will ever know if the Infomercial works.

These two elements, in tandem, are the most important and misunderstood elements in the whole process. Executives from the world of retail see to grasp the theory behind a successful media and telemarketing plan more so than executives from the media side. The success of an Infomercial has nothing to do with ratings or shares. It has to do with the "COST OF SALE". If you try to put it in any other terms, you will not only loose sleep at night, but you may also misinterpret the results of the media buy, and possibly pull the Infomercial off the air prematurely.

Let's take each element one at a time. First, let's discuss the media buy and how it should be effectively executed, tested and evaluated.

As with production companies, your choice of who will buy the media time on television should be limited to companies who specialize in the placement and buying of time for Infomercials.

As of this writing, there are only a handful of qualified media buying services placing Infomercials on a day to day basis, and some of these companies are also involves in the production of their own Infomercials. Which does not mean they are more or less qualified to produce Infomercials as opposed to production companies that are only in the production business?

However, many of the major advertising agencies who represent consumer marketers and Fortune 500 companies are looking into Infomercials. They are beginning to buy Infomercial media as well as producing Infomercials in-house and hiring the services of freelance Infomercial writers, producer, and directors.

Currently there are a small number of media/production companies in the Infomercial business. All of these companies have produced highly successful Infomercials as well as purchased their own air time on various cable networks and broadcast stations across the nation. What makes these media/production companies unique, is that they have purchased blocks of time on various cable television networks such as The Discovery Channel, and Lifetime Network, just to name a few. Having the foresight that the Infomercial business was about to become a booming business, they purchased large blocks of time from these cable networks in the early morning and late night hours that otherwise were generating little or no revenues for advertising for the stations in those tie slots.

In addition, the stations had to buy traditional programming for those time slots. The cable networks were happy to sell to these companies' blocks of none or little revenue generating time slots for a favorable cash price and at the same tie provide free programming to the stations eliminating their need to purchase other programming.

The result is that these companies have in their control, tie slots to air their Infomercials at a price that allows the not only to air their Infomercials, but to be able to sell time to other Infomercial producers and advertising agencies, yet still make a profit. As I mentioned, there are also a number of other companies who specialize in the placement and buying of media time for Infomercials. Again, some of these companies also produce their own Infomercials.

A few of those companies are Western media, Hawthorne Communications, Williams Television, and O2 Media. I mention these companies because they specialize in the media placement of Infomercials as well as producing Infomercials. On the other hand, production companies produce Infomercials and then place their media buys with a media buying agency

There are a number of ways to structure a deal with any of these companies for the placement and or production of your Infomercial. Of course, the more involved they become financially, the less profit participation you can expect.

Remember, the following scenario with respect to royalties only applies to the entrepreneur or manufacturer who wishes to sell product from a television offer directly to the consumer. This does not apply to the hybrid

Infomercial where the advertising agency pays a fee to the production company for the production.

The first scenario is to approach the with your product and try to strike an amicable deal whereby, they will finance the buying of the air tie, oversee and hire the inbound telemarketing company, and handle fulfillment of the orders.

The second scenario is to have the Infomercial produced yourself, or finance the production, hiring a reputable production company who does not buy media time. Then a deal can be struck with these companies to handle the media buys, inbound telemarketing, and fulfillment only!

Scenario number three is to have the Infomercial produced by an outside production company or one of these companies and for you to finance the buying of the time yourself, and purchase the time buys from the blocks of time they have available on stations or cable networks. You then can hire an in-bound telemarketing firm, as well as arrange for the fulfillment of orders from among hundreds of companies available throughout the country who specialize in fulfillment for direct response.

Although you will retain ore of the controls and profits in choosing this scenario, be prepared to have "deep pockets", and be prepared to hire people who have done this before.

Yes, this scenario can generate more profits, but the trade off is that although you ay not realize it, you have just formed your own direct response business and media business together with all the headaches associated.

Remember, all of these scenarios carry a price tag. The more these other companies are involved and exposed from a financial standpoint, the ore you will give up.

Chapter 11
MEDIA AND MEDIA COSTS

Where you air the Infomercial, what time of day or night you air it, and how frequently you air it, all are critical decisions in the testing process and evaluation of the Infomercial. There is ample data to determine the demographics of a station or cable network during different tie slots to target an effective media campaign. Some expects claim that an Infomercial can be tested by two or three airings on a small station in a small market where the media costs are low.

Others claim they can evaluate the Infomercial's potential success or failure by airing it on a cable network in the wee hours of the morning. For the most part, they are all correct. Even though the number of orders may be as little as ten or twenty from those airings, the experts can weigh the response against other successful Infomercials they have tested over the years on those same stations during the same time slots with similar products.

I have a basic problem with this type of evaluation, in that the test is missing the intangible factor of **"frequency"**. As with traditional TV commercial advertising, frequently and multiple impressions play a major role in the success or failure of a product. This is the basic problem I have with the traditional testing of Infomercials.

From time to time, we have all watched Infomercials, and if you watch television as a rule, you know we are bombarded with TV commercials selling thousands of different products.

We may not go out and buy a new car or box of cereal just because of a commercial. But if we were in the market for such a product, we would recall the television campaign, which in turn may impact our decision to purchase based on familiarity. This same theory can be applied to Infomercials in my opinion.

The consumer may not pick up a phone the first time he or she watches an Infomercial, but if they were to see it again and learn something new about the product they may have missed the first time, they might pick up the phone after watching subsequent airings.

The problem with this theory is that it can't be measured, and therefore can be debated for decades.

But the bottom line is that a media campaign to test the Infomercial can be accomplished with some certainty under conditions and parameters I mentioned above. If some frequency can be added in the number of airings, I feel the results of the test can be more accurate.

The way of measuring the success of an Infomercial is by the "COST OF ORDER". As a rule of thumb, media costs should not exceed fifty percent of the sale.

If your media costs are running twenty to thirty percent of your costs per order, you have a "sure fire" winner. It is prudent to test a product's potential by conducting a "test media buy" first, before spending sizable dollars in an all out campaign, or as it is called in the industry, a "roll out". This is usually where the misinterpretation begins for a newcomer to the world of Infomercials. Let's give an example of where the misconception begins.

You have a product or service that you are offering on an Infomercial. You test it on a Buffalo station at 1:00 AM on a Saturday night. You have just spent $100,000.00

on the production of an Infomercial, and want to see results. On Monday morning you've received only 100 orders.

Depressing, right? Wrong! Before you get depressed, you must analyze the "Cost Of Order".

Let's say that the selling price of the product is $60.00 plus shipping and handling of $4.00, bringing your total gross per order to $64.00. Your costs to manufacture or buy the product are $10.00. Your costs of inbound telemarketing are $2.00 (an average) for the operator to take the call, take the order, and process the credit cared information. Your cost to ship and handle the product order is $4.00. Your costs of buying the time on the Buffalo station are $400.00. That means your cost per order is $10.00 for product, plus $2.00 for inbound telemarketing, $4.00 shipping and handling and your media cost is $4.00 per order, for a total cost per order of $20.00.

Don't get depressed, you have what seems to be a sure winner. I hope this illustrates my point. It is not the number of orders, but the cost per order vs. media cost that matters when conducting a test.

Every sensible Infomercial producer or media buyer will conduct a media test similar to this. They will test in a number of different markets at the same time, and will air the show at different times of the day and night, as well as different days of the week. Many times they will also try different prices in different markets. This is the shotgun approach. However, as digital technology, the internet, and hundreds of cable and satellite networks emerge over the next decade, I predict that media buyers will be able to specifically reach demographically targeted audiences.

A good sized media buy for the test of an Infomercial should cost between $10,000.00 and $20,000.00. Once you have determined you have tested the product properly and can maintain a good "cost per order", you can "roll out" the Infomercial and purchase more expensive time on multiple stations that reach more potential buyers.

It is not unusual for a successful Infomercial to air between one hundred and one thousand times in one month. You may be asking yourself, what about the cost of shipping and handling the product. As part of the sale, the customer is charged shipping and handling fees on top of the actual selling price.

These fees should pay for any shipping and handling of the product by the fulfillment company. In many cases, there is a profit to be made from shipping and handling after the actual costs for same are deducted from the fee generated.

Shipping and handling charges can range from $3.95 to $14.95, depending on how much the product weighs and the handling and shipping costs involved. The customer expects it.

Most experts agree that the optimum time of year to air Infomercials are in the first and fourth quarters. Around mid second quarter, sales usually begin to diminish due to a number of reasons, mainly tax season, better weather, and people are watching less television.
On the other side of the coin, some of the best media rates for airing an Infomercial are in the second and third quarters.

To prove the point that first quarter is the best time to air an Infomercial, turn to any channel during the late night and early morning hours during the months of January, February, and March. You will probably see the air waves bombarded with a variety of INFOMERCIALS offering personal exercise equipment and some form of cosmetic, self improvement, or weight reduction system.

There is a simple psychology behind this category of product being offered more frequently during this time of year. It is during this period of the year when millions of people make their New Year's resolutions to eat better, feel better, and get into shape.

Chapter 12
THE ADVERTISING AGENCY

There are a growing number of advertising agencies that are beginning to utilize Infomercials for the consumer marketers they represent. As I mentioned in earlier portions of this book, these are hybrid/high concept Infomercials designed to introduce new product, create consumer awareness, enhance the corporate image, drive retail sales, create leads, or just give additional information about a product or service.

This is new territory for the advertising agency business and creative directors. It is also a new area that is unfamiliar in a business that has remained "status quo" for almost four decades relying on traditional spots, frequency, and imagery to create brand awareness and drive retail sales. Thus many agencies are approaching Infomercials reluctantly, while other agencies are listening to their clients and realizing that the agency business as a whole is a dinosaur and that the American public wants and needs more information and data before they make their purchasing decisions.

This is where the hybrid Infomercial I spoke about is leading. My philosophy is "THE MORE YOU TELL, THE MORE YOU SELL".

Those agency creative directors who are not averse to learning more about direct response television will realize that Infomercials are a new, innovative, fresh, and effective means of delivering their message to a shrinking and fragmented television audience. The other agencies who continue to resist this new form of television will lose more and more clients to the entrepreneurs such as CAA (Creative Artists Agency).

It is y opinion that the advertising community as a whole should take an aggressive and offensive approach to long form direct response television rather than a defensive (how do we keep everything status quo) approach.

The part that the advertising agency will play in the buying of media times for their clients is also changing. Many of today's advertising agencies are expanding into Infomercials to service their clients.

If you are a consumer marketer considering using Infomercials as an adjunct to an existing advertising or marketing campaign, you are probably using the services of an advertising agency for the creative and media buys.

Traditionally, it has been the advertising agencies that have brought the television media time to air their clients 30 to 60 second commercials, for which they have charged their client a flat percentage of the gross dollars spent. The media costs have always been divulged to their clients, and the fee for placement has always been above board.

One of the biggest complaints I have with the Infomercial production and media community is the lack of disclosure and transparency to their clients with respect to the cost of media. One of the best kept secrets in the Infomercial industry is how much is paid for time blocks on various cable networks.

If you, as an entrepreneur, consumer marketer, or advertising agency has produced an Infomercial or has had an Infomercial produced by an outside production company and wish to purchase time from one of these

Infomercial media companies on the cable network blocks they control, it is doubtful if you will ever receive a clear explanation of their cost of air time.

However, as more and more Fortune 500 companies and their powerful advertising agencies enter the world of direct response television and Infomercials, you will see a drastic change occur in the coming years. Until these agencies learn the Infomercial business and understand how to buy and place the media, they will hire the services of these few companies to buy and place Infomercials.

I predict that as major consumer marketers begin utilizing Infomercials as a means to introduce new product, create leads or drive retail sales, the advertising agencies will eventually control the Infomercial media blocks on cable networks and local broadcast stations just as they do now with commercial advertising spots. Just keep in mind that although these few Infomercial companies control certain cable network blocks, they cannot fill the time slots with their own Infomercials and product alone. They must sell time to other Infomercial producers and advertising agencies. You can negotiate the cost of your

Infomercial media with these companies. And if you cannot strike a satisfactory you can buy time on throughout the U.S... Also keep in mind that there is a built in conflict of interest when buying time from these few companies that control Infomercial time blocks.

Remember, they are also utilizing these time slots to promote and sell their own products on Infomercials they control. Make the specify what time slots you are buying, or when your Infomercials will air. If you don't and they have an Infomercial that is running and performing outstanding results, your Infomercial will get bumped to the wee hours of the morning where the viewing audience is greatly diminished.

Chapter 13

IN-BOUND TELEMARKETING

As with all of the other elements that go into the successful Infomercial, in-bound telemarketing is the final and probably the most crucial element. It is also the weakest link in the direct response chain. Ask anyone in the industry. This is the company or service that takes the calls from the consumer, answers any of their questions concerning the product, processes their credit card information, and turns that information into an order. There are hundreds of inbound and outbound telemarketing services that handle infomercials and DRTV. Do your research and search the internet.

DO NOT consider handling in-bound telemarketing yourself. The costs are too prohibitive. Hundreds of phone banks connected to Toll Free lines are housed at companies manned by hundreds of operators who handle not only your Infomercial orders, but possibly five or ten other Infomercial orders or mail order products. Their **only** information regarding the product and what is said to the consumer is called the "script". The "script" as it is called, is just that. On each operators computer terminal is the dialogue and information on the product they must relay to the customer when they phone in to order a product.

It includes all of the pertinent information such as price, product information, shipping and handling charges, mailing address, delivery dates, and "upsell" information, which I will go into a little later. If the script is not accurate, you can bet the order will not get processed properly, or worse yet, you will lose the order.

Remember, each operator is dealing with multiple products, and their only information is what is on their computer terminals. Whether you hire the services of an inbound telemarketing company yourself, or hire an Infomercial service such as the ones mentioned to do it for you, make sure the information given to the in-bound telemarketing operators is succinct and accurate.

Costs for processing calls can vary depending upon which company you select, how busy they are at the time, and how much time it takes for an operator to process an order.

The average cost of in-bound will range from a low of $1.50 per call to as high as $2.50 per call. The $1.75 to $2.50 charges are on a per call basis, and are charged for each call whether it translates into an order or not.

As each call is taken, the operator enters the information into a data base. The data base is then transmitted electronically or by hard copy to you, or the fulfillment company that will physically ship your order. You can monitor each day's sales by calling the company and giving them the private "access number" which is given to you when you contract with them to handle the in-bound telemarketing, or more recently, sales orders can be viewed and accessed on a secure internet website.

You can arrange for the fulfillment company to use their merchant credit card processing number in many cases, however, it is highly recommended that you have your own merchant account number.

In either case there is an additional charge for that service as well (approximately 3%). Payment is sent to you proof that the merchandise has been shipped. Remember, by Law the consumers credit card cannot be charged until the goods are actually shipped.

If you are dealing with one of the big companies in the Infomercial business they will usually contract with the in-bound telemarketing company and take care of the processing of orders and receiving the payment for the product. This, again, depends on the business deal you strike, and which one of the scenarios you choose from my previous illustrations.

There are, however, barometers to measure how effective the in-bound telemarketing service is. As a rule of thumb, a good percentage of "calls vs. orders" is 75%. That means for every hundred calls, 75% should translate into orders.

If the ratio is lower than this, you need to re-examine the "script" at the telemarketing level, or consider re-editing the Infomercial, because the consumer has not been given enough information or the "sell" portion of the script or "call to action" is not strong enough. Don't be afraid to call some of the people who ordered or didn't order. Give them a call and ask questions like why they decided not to order, or what prompted to pick up the phone to order, what did they like or dislike about the Infomercial, etc. This can be valuable information, the results of which can help in your evaluation of the Infomercial or possible changes that need to be made.

Now we come to the ""UPSELL"". An "upsell" is an additional product offered to a caller after the initial order has been processed. An example would be if you were offering a kitchen cooking utensil on an Infomercial.

After the operator has taken the order for the product offered on the Infomercial, he or she ay say to the caller, "thank you for your order! As an additional bonus, I am able to offer you today the "XYZ" cookbook for the low price of only $9.95 plus shipping and handling...may I add this to your order"? This is called an ""upsell"", and can generate additional income for the same amount of media buy. In essence, the "upsell" can lower your "cost per order" for no additional media cost.

Again, as a rule of thumb, if you have a product and decide to offer it as an "upsell", your ratio should be approximately 20% to 30%. That is for every confirmed order you receive, 20% to 30% of those buyers should buy the ""upsell"" product. The ""upsell"" can make the difference between a marginally successful Infomercial and successful Infomercial.

As with all elements that go into the successful Infomercial, carefully choose your in-bound telemarketing company. Sometimes it is prudent to go with a medium sized company as opposed to one of the large companies. That way you can receive a little more personalized service. For an extra fee, most of the in-bound telemarketing companies will offer a product orientation screening, where they screen the Infomercial for their operators and supervisors. This is very helpful for the operators, as it familiarizes them with the product and "offer". It is well worth the money. Remember, most operators never see the Infomercial; they merely go by the "script".

Whenever possible, ask to be present when the Infomercial first airs so that you can see how the operators are handling the calls, and make sure that the "script" is right. Remember, the in-bound telemarketing operators have no stake in the success or failure of your Infomercial, they are not on commission.

For the most part, they are average people who are supplementing their income, or students trying to make a few extra dollars part time. They are, however, the difference between making or breaking your success…stay on top of them!

Chapter 14
FULFILLMENT

This is the last step in the Infomercial and direct response process. Fulfillment encompasses the processing and data entry of the information supplied by the in-bound telemarketing company, plus the credit card approval, shipping, handling, returns, defects, and customer service aspects. Depending upon how many elements go into fulfillment of the order, your costs will vary. As an example, if your product is one product of the same size and weight, your costs for shipping and handling will remain consistent.

However, if the product is the type that varies in size, amount, and number, the cost to handle and package that product will be much more due to the additional manpower required.

Try to determine, in advance, what your product will cost to ship and handle before you produce the Infomercial. In most cases, the additional fee the consumer will pay for shipping and handling should cover your costs.

If there is an average cost of shipping and handling which is currently being used, it ranges between $3.95 and $4.95. On some Infomercials where the product is large and heavy, I have seen shipping and handling as high as $24.95. Plan accordingly.

Also remember, the Government has very strict guidelines when it comes to direct response. By Law, you must offer a thirty day money back guarantee. You also cannot charge the consumer's charge card until the order is hipped. Avoid C.O.D. orders. There is a high return rate, but you do not have to absorb the shipping and handling as long as somewhere in the infomercial or inbound telemarketing sales script you make the consumer aware.

Mail orders which encompass money orders and checks will account for about 10% to 12.5% of your orders if you decide to take those types of orders. There is also what is referred to as "lag orders". Lag orders are orders that come in two or three days, even weeks after an Infomercial has aired.

The theory is people may want to order your product after they have watched the Infomercial, so they write the phone number down and it sits on their coffee table for a day or so until they get around to calling and ordering.

Think twice about handling fulfillment yourself, unless your company is set up for it properly, or unless you want ulcers at an early age. It is a labor intensive endeavor, and does not stop with shipping the order. You must also maintain a customer service department to handle all complaints, returns, defects, credit card charge backs, and servicing, promptly for a period of time after the order is shipped.

Chapter 15

OUT-BOUND TELEMARKETING

This is another potential profit center that could add to your net profits. If you manufacture additional products, or have the ability to sell other products or continuing services, it is a very wise marketing strategy to include a brochure or advertisement when you ship the initial product the consumer has just ordered from your Infomercial. Why not! You have access to their address, telephone numbers, and credit card information.

If you can make an additional sale via the enclosed brochure, why not take the opportunity. You ay also wish to sell additional products via outbound telemarketing, whereby you hire a company to make calls to the persons on your mailing list to sell additional products.

Remember, the mailing list and information that you generate from the Infomercial is valuable property, and can translate into profits down the line. Protect it with your life. Each year hundreds of millions of dollars are spent purchasing mailing lists.

Chapter 16
THE FUTURE

The future of Infomercials as a means to sell product directly to the consumer and "drive" retail sales for new product is tremendous. I predict that over the next few years, major fortune 500 companies and their advertising agencies will look to hybrid Infomercials as a means to launch new products to the marketplace and consumer, and augment advertising campaigns to build product awareness.

There is no better costs effective method of informing and educating the public then with Infomercials, and they have proven time and time again to drive retail sales. I also predict that production values and quality of production will continue to increase. There will be no room for mediocrity with respect to production. Celebrities will demand more money and a bigger piece of the profits, and broadcasters, cable networks, and local cable stations will raise their price on media time.

More and more production companies will pop up to answer the demands for Infomercial production. The companies producing Infomercials will also be segmented into two factions. There will be the production and media companies who have a royalty position on every product sold over television and who will always be looking for the next exercise, beauty, or fitness product to be sold primarily through direct response.

There will also be the production company who will produce these hybrid Infomercials for Fortune 500 companies and their advertising agencies for only their production fee, much like the commercial production companies now serving the agencies.

I also foresee heavier Government restrictions on the Infomercial industry and the weeding out of those who deceive the public with false claims and inferior products. In answer to the mounting pressure from lawmakers and government officials critical of the Infomercial industry, a small group of the key players in the field of Infomercials formed the National Infomercial Marketing Association in 1990.

Based in Washington, D.C., this organization is comprised of over 200 companies providing services to the Infomercial industry. Their goal is to self-police the industry before Government does it for them.

They have met with Congressional members and Federal Trade Commission officials to set certain guidelines for producer of Infomercials to follow.

These guidelines reflect NIMA's belief that truthful and informative programming is the best way to win and hold public confidence and gain lasting acceptance and success for the Infomercial industry. One of the main requirements of each NIMA member is that Infomercials which they produce be clearly identified as "paid advertisements" on the screen at the beginning and end of each aired program, as well as in the body of the program when specific ordering information is given to the consumer.

I also foresee in the very near future, the start up of a new 24 hour Talk/Infomercial cable network, which will become the information and retail store of television.
I predict the Home Shopping Network and QVC will also expand their formats to include long form Infomercials and develop new forms or direct response and interactive programming.

New digital technologies such as addressable transmissions will enable cable systems and home shopping shows and networks to transmit Infomercials to a specific local cable channel in the middle of a major city, where desirable demographics are targeted for a specific product.

You may also see one or more of the major networks program a hybrid Infomercial/talk format in the early morning hours which will be entertaining, informative, and sell product through an 800 number. You will also see an expansion of products offered on Infomercials.

Manufacturers will use the INFORMERCIAL to introduce new products to the public. An example would be Ford or Chrysler producing INFOMERCIALS to launch a new line of automobiles.

Not with the intention of selling cars over the television, but to inform the public and drive retail sales.

Time slots will also become harder to secure with the overabundance of Infomercials produced over the next few years, all needing air time. In fact, as a producer of entertainment first run syndicated programming, I predict it will be almost impossible to get on the air with a new series in early morning or late night fringe or weekends. However, many new cable networks will emerge in need of infomercial dollars for their 24/7 programming schedule.

With Infomercial producers and media buyers offering cold hard cash to stations for the airing of Infomercials, what is the incentive for stations to try a new show on a barter basis that has to find an audience, get a reasonable rating, and be sold to sponsors.

And don't think that the Fortune 500 company's advertising agencies are just going to sit around and let someone else control the production and media buys. I predict that in the next two years, advertising agencies will rule the Infomercial production and media buying as they do now with print and traditional commercial advertising.

I am also predicting Infomercials will eventually air primarily on cable stations and cable networks, due to broadcast television's "love/hate" relationship with Infomercials and their demand for higher rates, putting the "cost per order" at an unreasonable ratio.

Finally, from the convergence of broadcast, cable, and the internet, a new 24/7 network will evolve with "live" and "on-demand" long form infomercials and short form DRTV spots offering a wide range of products and services through using highly produced HD streaming video productions, that will encompass payment, shipping, and fulfillment from one website.

Chapter 17
INFOMERCIALS SCAMS

There are many great advertising and direct response agencies throughout America, agencies created and managed by people with a strong sense of right and wrong and focused on holding our profession to the highest standards.

Unfortunately, like in most industries, there are also those "agencies" that are in it for the quick buck. These people will take money from unsuspecting businesspeople and entrepreneurs, fully aware that what they are offering is of no value.

These scam "agencies" offer a variety of different "amazing" advertising opportunities, but they have one thing in common: you lose your money, and they get wrongfully enriched. The following is a brief description of the most popular scams, how they operate and how you can uncover their real intent.

THE TV SHOW SCAM

There are several so-called show producers working this scam (many of them right here in Florida). It goes like this: you are contacted by a TV show "producer" who tells you that he produces a weekly or monthly TV show; he tells you the show focuses on lifestyle subjects such as home improvement, decorating and so on.

Lucky for you, they want to do a segment on a product just like yours. Even more exciting, the show is hosted by some "C" level celebrity and will air all over the country on major TV networks like HGTV, Lifetime, TLC and others. All you have to do is cover the "cost" of producing your segment, usually between $5,800 and $30,000. Finally, you will need to make a decision right away since either some other company is chomping at the bit, or the segment is about to go into production and they must know if you are in or out.

Let's take a look at what is really going on here. First of all, if a real TV producer were to contact you and ask to cover your product or service, they would not ask you to pay for the segment. Many legitimate shows feature companies and their products every day, but the producers make decisions on what they feel fits the show and is newsworthy, not who will pay to appear in the show.

How about all those big national TV networks the "show" will air on? Well, they do air it on networks like HGTV, but do not be fooled: it is not aired nationally. Instead, the "producers" buy out or pre-empt a half-hour block of time on a network in a small local market or two. So they do indeed air it on a network, but it is in a very small market where only a few people actually see the broadcast.

THE DRTV SPOT SCAM

As in the above scenario, in the DRTV Scam someone who wants information about your product contacts you. He tells you that your product would be great for direct response (DR). He also tells you that, not only does he think it is a hot product; his company is willing to "invest" some of its own money. What could possibly go wrong in this situation?

The representative will tell you that they will produce a commercial worth roughly $25,000 and run a schedule of hundreds of commercials on major networks. All you must do is put up part of the money (usually a few thousand dollars) and, if the product does well, they get a small percentage of sales. The problem is this: there will never be any sales.

What you really get is a cheap, badly-produced commercial that is worth nearly nothing, which gets aired on some major networks, again in a couple small cable markets. And again, if you are lucky, a couple of hundred people might see it.

There is also a version of this scam that revolves around an infomercial, but it plays out much the same: invest a few thousand dollars and your trust with a company and let them run with it. The result is the same, too, in that you lose the money you invested and you're left a little poorer with a bad view of "advertising agencies."

THE SHOPPING NETWORK SCAM

First you get a call from a product "buyer" that works with one of the shopping channels or shows. This buyer says that he wants to buy some of your product to place on his show and, if your product sells out, he will buy a lot more of it. The offer sounds good – until the other shoe drops.

As the pitch goes on, you find out that he will buy $400-$500 worth of your product, but you will need to put up between $5,000 and $20,000 for the original show production. Of course, if it sells out, he will be buying "massive" amounts of your product.

What will really happen is you get a segment in a badly-produced shopping show that, you guessed it, airs on some TV network in an obscure, small market. Not surprisingly, the product never sells out.

IDENTIFYING A SCAM

There are numerous variations on these scams, but the net result is always the same: you lose your valuable marketing money and get nothing in return. Since so much is at stake, always be wary and remember that anything that seems too good to be true probably is just that.

Beyond wariness, use this simple, surefire method to sniff out the fakes: ask the "producer / buyer /

representative" (or whatever else they may call themselves) for the names and phone numbers of three current and successful clients that are on the air.

At this point you will hear one or more of the following: A) The producer/buyer/rep is new to the firm and doesn't know the client list yet; B) The information is confidential, and they can't give it out; C) He will give you the name of some well-known products, claim his company handles them, but he can't give you the names of the contact people; D) He promises to get the information and call you back, but you never hear from him again.

If you hear any one of these or a variation thereof, walk away from the deal. If the producer / buyer / representative are at all wishy-washy, walk away from the deal. Most important, if your gut tells you something seems fishy, walk away from the deal.

Our goal here at the **CUMMINGS ENTERTAINMENT GROUP** isn't to dissuade you from doing business with someone else; it is to offer people the best advice and recommendations we can provide. Sometimes what we have to say is not what people want to hear, but we owe everyone who comes in contact with us the truth.

Ultimately, that comes down to one final point. There are many legitimate, well-operated DRTV production companies, advertising agencies, and media buying companies in this country. Speak to them, listen and choose someone that you feel will work in your best interest. Although there are no guarantees in the business of Direct Response marketing, working with a real professional is your best chance at success.